ISBN 9780648608707

To Georgia and Scarlett,

May your brains be full of knowledge and your faces full of smiles.

Children are remarkable for their ability to quickly and easily learn massive amounts of new information. It is proposed from recent research that the key behind this ability to rapidly process is the neurotransmitter GABA. Furthermore, functional MRI have proven children activate different and more regions of their brains than adults when they perform word tasks*.

*Washington University School of Medicine in St. Louis.

Why is it then if children's brains absorb information at a much faster rate than adults that they are so hard to teach at home? We have a theory...

"The average 4-year-old laughs 300 times a day. The average 40-year-old? Only four."

This common quote is hard to attribute to a single individual and trace its origin. However It is undeniable to a parent that children laugh longer and louder and are more distracted than adults.

The desire to have fun competes with their increased capability to learn. Imagine a world where you both win. They do their homework (tick parenting win) and they have fun (tick child win). This lead to our research and development of Learning through fun.

This volume focuses on mathematics and is based upon the New Zealand Curriculum, Australian Schools Curriculum (Version 9) and Naplan testing 2022. Each book explores visual and non visual mathematics that when answered will allow the preparation of recipes. There will be times when they will need guidance.

These recipes are predominantly healthy with a scattering of special treat recipes included. Nut and other common allergens are absent from the majority of sandwich fillings and school snacks however we ask that you review all recipes carefully if you have an allergy.

INDEX

SWEET BISCUITS

Peanut Butter Choc Chip Cookies

STEP 1

Pre-heat the oven to 200-10 = °C

STEP 2

Cream together in a bowl

BUTTER softened

= cup

BROWN SUGAR *6 x ? =6* = cup

STEP 3

Mix until combined

EGG *4 apples take away 3 apples* =

Then add

PEANUT BUTTER one  cut into 2 = a  + a ? = cup

VANILLA EXTRACT

How many whole apples is  +  +  +  ? = tsp

Mix well.

STEP 4

Whisk the following dry ingredients together then add to the butter mixture, mixing together thoroughly.

PLAIN FLOUR ¾ 1 ? 1½ = cups

BAKING SODA *1 – ½* = tsp

SALT *2 X ? =1* = tsp

STEP 5

Evenly mix in

CHOCOLATE CHIPS 8 6 4 ? = cups

STEP 6

Using an ice cream scoop or rounded desert spoon. Place rounded cookie scoops onto lined baking sheets.

STEP 7

Bake in the oven until golden brown all over.

TIME *5+5+5* = minutes

STEP 8

Remove from the oven. The cookies should still be in a dome shape.
Using a flat spatula press each cookie down until evenly flat. They should be about 1.5cm thick.

STEP 9

Leave the cookies on the trays to cool completely after flattening.

Coconut Meringue Cookies

STEP 1 Preheat oven and line two baking trays with baking paper.

TEMPERATURE *100 +60* = °C

*Add 20°C if not a fan forced oven

STEP 2 Whip

EGG WHITES  *take away 5 oranges* =

CREAM OF TARTAR *¼ x 1* = tsp

CASTER SUGAR

Write a fraction where the numerator is 3 and the denominator is 4. = cup

STEP 3 Fold in

COCONUT FLAKES *15-14* = cup

CORNFLAKE CRUMBS (CRUSHED)

What fraction of the pizza is coloured? 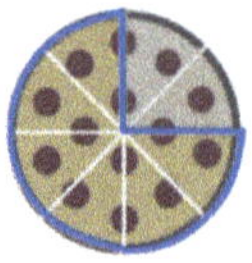 = cup

STEP 4 Drop by spoonful onto parchment lined cookie sheets.

STEP 5 Bake until golden.

TIME *6 + 6 + 6* = minutes

SAVOURY BISCUITS

Seed Cracker

STEP 1

Place all ingredients in a bowl and mix. It will be quite wet

FLOUR (almond or wheat)	*? + ½ =1*	=	cup
PEPITA/PUMPKIN SEEDS			
	In the fraction 1/16 the numerator is ?	=	cup
CHIA SEED	*¾ +? =1*	=	cup
SEA SALT	*12 -? = 11*	=	tsp
OLIVE OIL	*6 ÷ ? = 3*	=	Tbsp
WATER	$\begin{array}{r} 150 \\ +\ 100 \\ \hline = \end{array}$	=	ml

STEP 2

Let sit 10 minutes, just until it starts to bind together. During this time line a baking tray with greased baking paper and

PREHEAT OVEN	*18 x 10*	=	°C

STEP 3

Evenly spread out mixture on baking paper. Then take a metal spatula dipped in water and smooth the mixture.

THICKNESS	$\begin{array}{r} 16 \\ -\ \square \\ \hline 13 \end{array}$	=	mm

STEP 5

Gently score the mixture. Bake for

TIME	*¼ of an 4 hours*	=	hour

STEP 6

When slightly cooled break the crackers and turn over on the cookie sheet. Slide back into the oven with oven temp turned off. Leave crackers in the cooling oven for another 15-20 minutes or until crisp through. They will crisp more as they cool.

SWEET MUFFINS

Cinnamon Apple Muffin

<u>STEP 1</u> Preheat oven and spray 12 muffin tin with oil or line with patty pans

TEMPERATURE	$\begin{array}{r} + \ 80 \\ 100 \\ \hline = \end{array}$	=	°C

<u>STEP 2</u>

Sift together in a large bowl

PLAIN FLOUR	*4 half pizzas = how many whole pizzas?*	=	cups
BAKING POWDER	*½ + ½ + ½*	=	tsp
CINNAMON			
Felix and Maggie equally share 5 apples. How much do they each get?		=	tsp
SALT	*2+2-3*	=	pinch

<u>STEP 3</u>

Dice apple into 1cm cubes and add to mixture

PINK LADY APPLES	*4+5 = 7+?*	=	cups

STEP 4

Cream together in a separate large bowl

BUTTER, room temperature	*1 apple cut in 2 = 2 x ?*	=	cup
CASTER SUGAR	 *?/3 are bananas*	=	cups
Next add one **EGG** at a time	*What is the numerator in 2/5?*	=	eggs
Mix in **VANILLA**	*Finish the sequence 14...10...6...?*	=	tsp

STEP 5

Gently fold in flour/apple mixture, alternating with milk. Stir until just combined.

MILK	*5 is what fraction of 10?*	=	cup

STEP 6

Fill muffin tins until ⅔ full. Bake until a toothpick comes out clean,

TIME	*Half of one hour*	=	minutes

STEP 7

Mix together sugar and cinnamon in a separate bowl

CINNAMON	*6x0 +3*	=	Tbsp
SUGAR 	*What fraction of the jug is filled?*	=	cup

STEP 8

Once muffins have slightly cooled, brush with melted butter and sprinkle with the cinnamon mixture. Alternatively you can dip the muffin into butter then cinnamon mix.

BUTTER, Melted	*¼ x 1*	=	cup

Banana & Choc Chip Muffin

STEP 1

Preheat oven and spray 12 muffin tin with oil or line with patty pans

TEMPERATURE If fan forced SUBTRACT 20°C	100 x 2 =	=	°C

STEP 2

Sift flour and baking powder into a large bowl. Stir in sugar.

WHOLEMEAL FLOUR	*10 ÷ 5*	=	cups
BAKING POWDER	*17-14*	=	tsp
BROWN SUGAR	*1- ⅔*	=	cup

STEP 3

In a separate bowl, mix together egg, oil and milk.

EGGS	*2 x 2 = 3+?*	=	
OLIVE OIL (Hint: Count pizzas not slices)	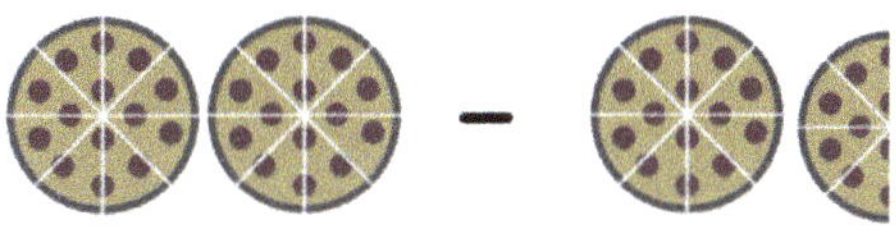	=	cup

MILK	*4 + 6 - 9*	=	cup

STEP 4

Add wet ingredients, banana and choc chips to dry ingredients and fold through gently.

BANANA, ripe mashed	76 - 75 =	=	cup
CHOC CHIPS	*8 is ? of 16 (clue ¼, ⅓ or ½)*	=	cup

STEP 5

Line muffin tin and spoon mixture evenly into muffin holes.

STEP 6

Bake for 20 minutes or until a skewer comes out clean.

NOTES

Chocolate Avocado Fudge Muffin

STEP 1 Pre-heat oven and line muffin tin

TEMPERATURE			
Finish the series	*100..120..140..?*	=	°C

STEP 2 In a large bowl combine the following ingredients

FLOUR	*? x 5 = 10*	=	cups
SUGAR	*¼ + ¼ + ¼*	=	cup
COCOA	*¼ + ¼*	=	cup
SALT	*1 - ½*	=	tsp
BAKING POWDER	6 - 3 - 2	=	tsp
Mix in **CHOCOLATE CHIPS**	*¼ x 3*	=	cup

STEP 3 In a medium bowl whisk the following ingredients

MASHED AVOCADO	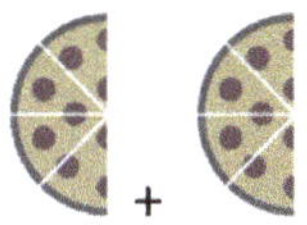	=	
GREEK YOGHURT	*50 + 50 + 50*	=	gram
SKIM MILK	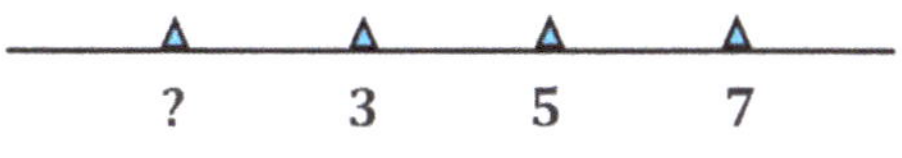	=	cup
COCONUT OIL			
If one pizza is split between 4 boys each will get what fraction?		=	cup
VANILLA	*4 + 3 - 6*	=	tsp

STEP 4 Combine wet and dry ingredients and stir until combined
Spoon into muffin tins.

STEP 5 Bake

TIME	20 x 2	=	minutes

SAVOURY MUFFINS

Feta & Roast Capsicum Muffin

STEP 1

Preheat oven and spray a 12 muffin tin with oil or line with patty pans

TEMPERATURE *How many cm in 2metres ?* = °C

If fan forced SUBTRACT 20°C

STEP 2

Sift flour and baking powder into a large bowl. Stir in sugar.

WHOLEMEAL FLOUR *? X 2 = 4* = cups

BAKING POWDER *10 - 7* = tsp

STEP 3

In a separate bowl, mix together egg, oil and milk.

EGGS *3 +? = 2 x 2* =

OLIVE OIL  *= what fraction of*  = cup

MILK *3X4 = 11+?* = cup

STEP 4

Add wet ingredients, capsicum and feta to dry ingredients and fold through gently.

ROAST CAPSICUM, chopped

What is the fraction if the numerator is 2 and dominator is 3?		=	cup
FETA, crumbled	*¼ + ¼*	=	cup

STEP 5

Line muffin tin and spoon mixture evenly into muffin holes.

STEP 6

Bake for until a skewer comes out clean.

COOKING TIME	*16 + 4*	=	minutes

Savoury Ricotta Muffin

****There is no flour in the recipe! It is not a mistake**

STEP 1

Preheat oven Line muffin tin with patty pans or spray with oil

TEMPERATURE *200-10* = °C

STEP 2

Slice olives, sundried tomatoes and place with capers in a large mixing bowl.

SUN DRIED TOMATOES *How many quarters make a whole?* =

OLIVES *John gave me ½ of his 4 apples. How many did I receive?* = Tbsp

CAPERS *20 cherries can be split into how many groups of 10?* = tsp

STEP 3

Add eggs, ricotta, spring onion, salt and pepper. Mix to combine.

EGGS *What is the difference in time between the 2 clocks in hours?*

 **And**  =

RICOTTA	*2 x 6 = 13 - ?*	=	cup
SPRING ONION	*3 x 6 =* 20 - ?	=	
SALT & PEPPER	*? x 6 = 7 - 1*	=	pinch

STEP 4

Fill each muffin hole ⅔ full

STEP 5

Bake in oven.

TIME	*30-10*	=	minutes

Notes

SANDWICH FILLINGS

Omelette Sandwich

STEP 1

Heat oil in a small non-stick frying pan and fry sliced capsicum and onion over a low heat until it is browned and tender, this will probably take about 10 minutes in all. Stir in the spinach.

OIL 	If I eat ¾ of the apples, how many remain?	=	tsp
CAPSICUM, thinly sliced	*1000÷10*	=	grams
RED ONION, thinly sliced	*14 is what fraction of 28?*	=	
BABY SPINACH LEAVES	*7X0+1*	=	handful
BABY TOMATOES, sliced in half	*3x* 	=	

STEP 2

Beat eggs and milk in a medium bowl. Pour mixture into frying pan and gently fold

EGGS		=	
FULL CREAM MILK	*A dozen = 10+?*	=	Tbsp

STEP 3

Cook the omelette mixture until edges are golden and crispy

STEP 4

Toast the bread and fill with omelette mixture. Should be enough for 2 sandwiches. If making for child's lunchbox allow toast and egg mixture to cool

White Bean & Cheese Sandwich

STEP 1

In a medium mixing bowl, combine beans, olive oil, salt, pepper and lemon juice.

CANNELINI BEANS, **washed and drained**	*½ + ½*	=	tin
OLIVE OIL	*15 x ? = 3 x 5*	=	Tbsp
LEMON JUICE	*90 minutes is how many hours?*	=	Tbsp
SALT	*6 x ? = 3+3*	=	pinch
PEPPER	*4 x ¼*	=	pinch

STEP 2

Roughly mash with a fork leaving a little chunky

STEP 3

Spread mixture thickly along slice of bread, sprinkle feta cheese and layer with lettuce or mint. If you like you can add crushed chili for a hit .

FETA CHEESE	*¾ + ? =1*	=	cup

LUNCH FOR THOSE WHO DON'T WANT A SANDWICH

Apple Weetbix Pikelets

STEP 1

Mix dry ingredients

WHOLEMEAL FLOUR	*10+10-19*	=	cup
CRUSHED WEETBIX	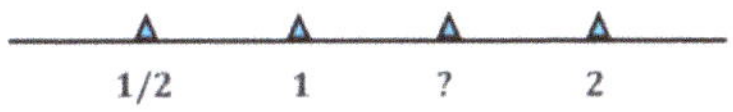	=	cups
BAKING POWDER	*2+6 =4 x ?*	=	tsp
CINNAMON	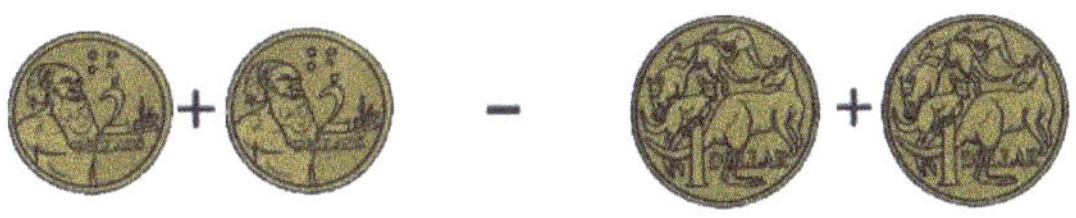	=	tsp

STEP 2

Mix in wet ingredients

EGG	*½+½ =2-?*	=	
BUTTER, melted	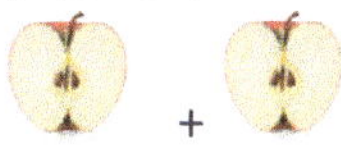 + 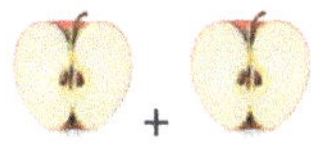	=	Tbsp
MILK	*½ + ? = 2*	=	cups

STEP 3

Heat frying pan over medium heat. Melt a little butter in the pan and dollop tablespoons of mixture into the pan

STEP 4

Turn when brown on the bottom and brown on the other side. Serve with your favourite topping or allow to cool and add to school lunches.

Tuna Slice

**Freezes really well and can easily be popped into lunch boxes

STEP 1

Pre-heat oven **TEMPERATURE**	80 + 100 =	=	°C

STEP 2

In a blender add biscuits, onion and cheese. Blend until coarse crumbled mixture.

DRY CRACKERS such as jatz	*½ + ? =1*	=	packet
BROWN ONION	*¾+¼*	=	
CHEDDAR CHEESE	*2- ½*	=	cups

STEP 3

In a large bowl combine eggs, milk, tuna, vegetables and cracker mixture. Stir thoroughly

EGGS	*3x3 = ?+3*	=	
MILK	*50 = ? OF 100 (hint ¼, ⅓, ½)*	=	cup
TUNA	*18+? =20*	=	tins
FROZEN VEGETABLES (carrot, corn and cauliflower)	*3x? = 4+2*	=	cups
ZUCCHINI, grated and fluid removed	*6+? = 10-3*	=	

STEP 4

Pour into shallow tray and Bake until golden.

TIME	10 20 ? 40	=	minutes

SLOW COOKER & ONE POT RECIPES

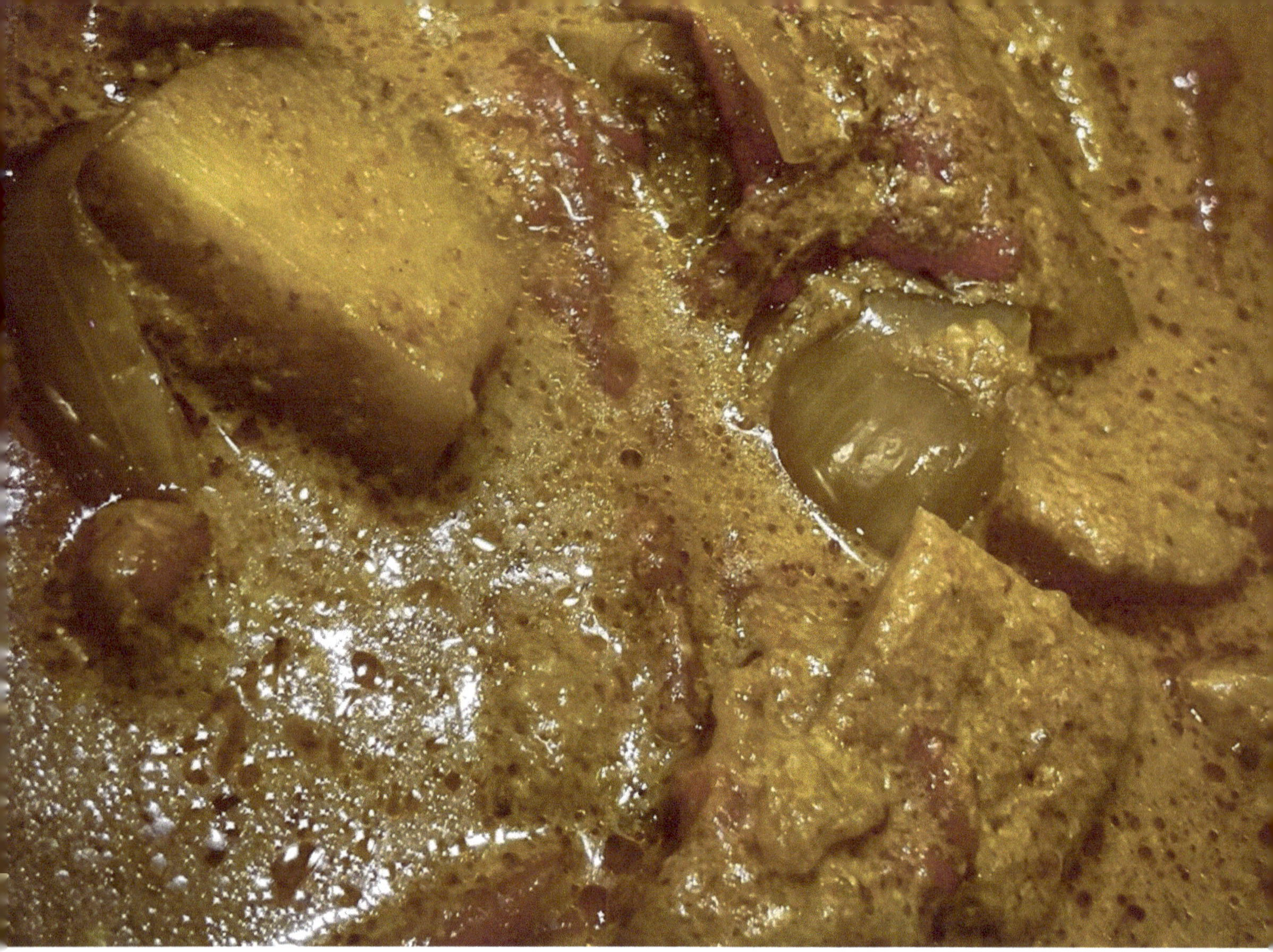

Slow Cooker Yellow Beef Curry

STEP 1

In slow cooker add the following ingredients

BEEF (Chuck, beef cheek) **diced**	*50x10*	=	grams
POTATOES, diced	number line: 1 ? 7 10	=	
ONION, diced	*17 + ? = 6 x 3*	=	
RED CAPSICUM, diced	*2+4+? = 7*	=	
COCONUT CREAM	$\begin{array}{r} 600 \\ -\ 200 \\ \hline = \end{array}$	=	ml can
BEEF STOCK	*5 = 4+ ½ +?*	=	cup
GINGER	*¼ + ¾*	=	Tbsp
YELLOW THAI CURRY PASTE	*9-6*	=	Tbsp
GARLIC, crushed	$\begin{array}{r} 12 \\ -\ 9 \\ \hline = \end{array}$	=	tsp

FISH SAUCE *How many more oranges to make a dozen?* = Tbsp

BROWN SUGAR *10x2= 19+?* = Tbsp

STEP 2

Cook on slow

TIME = hours

Jack has 16 cards. Ian has as many as jack. Mary has half as many as Ian. How many cards does Mary have?

STEP 3

In a small cup add corn flour and ladle of juice from curry. When combined pour mixture into curry and cook on high

CORN FLOUR 20+7-26 = Tbsp

CURRY SAUCE $\begin{array}{r}4\\ ?\overline{)8}\end{array}$ = Tbsp

TIME *5x6* = minutes

STEP 4

Before serving stir in lemon juice

JUICE *5x6 -29* = Tbsp

Tuna and Vegetable One Pot Dump Bake

The ultimate recipe when you run out of time

STEP 1

Pre-heat oven

TEMPERATURE	*2x50 + 2x40*	=	°C

STEP 2

In a large ovenproof dish combine all ingredients

BROWN ONION, finely sliced	*¼ of 4*	=	
FROZEN BROCCOLI, CAULIFLOWER AND CARROT MIX, thawed	*¼ x4*	=	kg bag
SHREDDED CHEESE	*10÷5*	=	cups
TUNA CAN, large	200 + 225 =	=	tin
CONDENSED CREAM OF MUSHROOM SOUP	*4x100 + 4x5*	=	gram can

SOUR CREAM

After I eat the selected portion what fraction remains? = cup

BLACK PEPPER $\frac{1}{4} + \frac{1}{4}$ = tsp

STEP 3

Bake covered with foil

TIME

 *and*  *What is the difference in time between the clocks?* = minutes

STEP 4

Remove foil and bake until golden

TIME $2\overline{)10}$ = minutes

Saag Paneer

STEP 1

Defrost frozen spinach and squeeze water. Set aside

STEP 2

Purée half of spinach in blender.

SPINACH	*17 – (2X8)*	=	packet

STEP 3

Add paneer to saucepan or deep frying pan. Toast until brown then set aside

PANEER, cubed	*How many cm in 6m?*	=	grams

STEP 4

Add cumin and olive oil

CUMIN SEEDS	*1+2+3+4-9*	=	tsp
OLIVE OIL	*½ of 2*	=	Tbsp

STEP 5

Continue by adding onion and garlic and sauté until onions are soft.

TIME	*Difference in time*  and 		=	minutes
GARLIC, minced	*1/3 of* 		=	cloves
RED ONION	$\begin{array}{r} 16 \\ -\ ? \\ \hline =\ 15 \end{array}$		=	cup
CANOLA OIL	*178+2-179*		=	Tbsp

STEP 6

Add pureed tomato and salt.

PUREED TOMATO	*2+12-13*	=	tin
SALT	*(½ of 4) -1*	=	tsp
COOKING TIME	*40÷8*	=	minutes

STEP 7

Add pureed spinach, Curry powder, coriander powder, chilli powder and water. Bring to boil.

WATER	*¼ + ? = ½*	=	cup
CURRY POWDER	*7___5____3____?*	=	Tbsp
CORIANDER POWDER	*¼ X 4 =*	=	tsp
CHILI POWDER	*¼ + ¼ + ? = 2*	=	tsp

STEP 8

Remove from heat and add butter, cream and paneer.

CREAM	8÷? = 4	=	Tbsp
BUTTER	?+5 = 2 x3	=	Tbsp
COOKING TIME	$\begin{array}{r} 78 \\ -\ ? \\ \hline =\ 73 \end{array}$	=	minutes

VEGETABLE SIDES

Shredded Brussels Sprouts with Pistachios, Cranberries & Parmesan

STEP 1

Cut Brussel sprouts in half through the root, then slice thinly. Carefully use a mandolin with an adult

STEP 2

Heat olive oil in a large skillet set over medium heat. Add the onions and cook, stirring occasionally, until the onions are soft, 4 to 5 minutes.

BROWN ONION, finely diced	*2+3-4*	=	
COOKING TIME	35 25 15 ?	=	minutes
OLIVE OIL	*2x2-3*	=	tsp

STEP 3

Add further olive oil to the skillet, then add the Brussels sprouts. Cook, stirring occasionally, until the Brussels sprouts are tender, but still bright green.

OLIVE OIL	*4x4 = 15+?*	=	tsp
BRUSSEL SPROUTS			
If I catch 100 balls in 1 hour how many will I catch in 5 hours?		=	grams

STEP 4

Stir in the pistachios, dried cranberries, Parmesan cheese and pepper

PISTACHIO	*What is the selected fraction?*	=	cup
DRIED CRANBERRIES	*? + ⅔ = 1*	=	cup
PARMESAN CHEESE	*1/3 + 1/3 + ? =1*	=	cup
PEPPER	*½ +2+ ½ + ? = 4*	=	tsp

STEP 5

Transfer to a serving dish and keep warm until ready to serve.

Homemade Pesto Pasta / Pasta Salad

A delicious dinner that also can be served cold as a non sandwich lunch option for kids. Use wholemeal fusilli, penne and spiralli when possible.

STEP 1 Cook pasta according to package directions.

SPIRAL PASTA	$\begin{array}{r} 47 \\ -\ 45 \\ \hline = \end{array}$	=	cups

STEP 2 In the bowl of a food processor, add basil leaves and pine nuts and pulse until very finely minced.

BASIL LEAVES, fresh	*16-14*	=	cups
PINE NUTS	*? x 16 = 8*	=	cup

STEP 3

Add the garlic and parmesan cheese and pulse a few more times until combined.

GARLIC	*11+? = 4x3*	=	clove
PARMESAN CHEESE	*3 = ? of 6 (hint: ¼, ⅓, ½)*	=	cup

STEP 4

With the food processor running, slowly pour in the olive oil and blend until combined and smooth. Season with salt and pepper to taste and pulse a couple of times.

OLIVE OIL	*2 is what fraction of 8?*	=	cup
SALT	*2 is what fraction of 4?*	=	tsp
PEPPER	*4 X ? =2*	=	tsp

STEP 5

Serve with cooked pasta with an extra sprinkle of parmesan cheese as a garnish. Add tomatoes and stir to mix. Feel free to mix in steamed vegies for extra nutritional value!

SUNDRIED TOMATOES, chopped	*½ + ½ + ½ + ? = 2*	=	cup

Photo: Lily Cowan, Year 8, Unley High School, South Australia

Pumpkin, Roasted Hazelnuts and Feta Salad

STEP 1

Preheat oven

TEMPERATURE	*100+50+50*	=	°C.

Place hazelnuts on a lined baking tray

COOKING TIME	*90 + ? =100*	=	minutes
HAZELNUT	*12 - ? - 3 = 8*	=	cup

STEP 2

Transfer hazelnuts to a clean towel and rub to remove skin set a side

STEP 3

Line large roasting pan with nonstick paper. Arrange pumpkin in and spray with olive oil. Season with salt and pepper

PUMPKIN cut into 3cm cubes	*1500g is how many kilograms?*	=	kg
SALT	*3x4 - 11*	=	tsp

PEPPER	*2+6-5 = 2+?*	=	tsp
TIME	*¼ of an hour*	=	minutes

STEP 4

Place the following in a screwtop jar. With the lid off microwave for 10 seconds. Replace lid and shake to mix.

HONEY 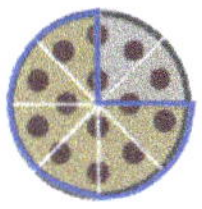	*How much pizza is left if I eat the coloured portion?*	=	cup
OLIVE OIL	26 18 10 ?	=	Tbsp
BALSAMIC VINEGAR	108 - ? =106	=	Tbsp

STEP 5

Combine pumpkin, hazelnuts and feta on serving dish. Pour over dressing

FETA CHEESE, **crumbled**	*100 +50+25*	=	grams

Notes

CHEESECAKE

Blueberry Cheesecake

STEP 1

Prepare cake tin by greasing and lining with greaseproof / baking paper.

STEP 2

Blitz biscuits in a food processor

DIGESTIVE BISCUITS	$\begin{array}{r} + 100 \\ 80 \\ \hline = \end{array}$	=	grams
TIME	*½ of a minute*	=	seconds

STEP 3

Melt

UNSALTED BUTTER	*20x4*	=	grams

Stir the crumbed biscuits into the melted butter.

STEP 4

Press biscuit mixture into cake tin, making sure bottom is covered evenly. Place in fridge for

TIME	*2x5 +10*	=	minutes

STEP 5

Whisk together the following in a large bowl

CREAM CHEESE	*50 x 6*	=	grams

DOUBLE CREAM	*50x6 -50*	=	ml
LEMON JUICE	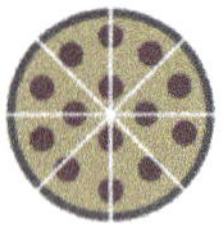 *How many slices make up half of this pizza?*	=	Tbsp
VANILLA EXTRACT	*15÷5 -2*	=	tsp

STEP 6

Pour the cheesecake mixture over the biscuit base and smooth cheesecake out.

STEP 7

Place the cheesecake into the fridge to chill for at least 4 hours or overnight.

STEP 8

Transfer from tin to serving dish, top with the blueberry jam and serve

OTHER CAKE

Peach Dump Cake

STEP 1

Preheat oven and grease a 9x13 inch baking tin.

TEMPERATURE	*200-20*	=	°C

STEP 2

PEACHES in syrup			
I have 4 cups of water and I pour out ¾ of them how many cups are left?		=	x 825g can
SUGAR	*3- 2 ½ =*	=	cup
CORNFLOUR	? 5 7½ 10	=	Tbsp
SALT	*5 = 4+?*	=	pinch

STEP 2

In a large bowl mix following ingredients

PLAIN FLOUR	*100 +50+25*	=	grams
SUGAR	*400÷2*	=	grams
BAKING POWDER	*47-37-9*	=	tsp
SALT	*1 - ¾*	=	tsp

STEP 3

Sprinkle mixture over the evenly spread fruit

Drizzle butter over top

BUTTER, melted	*17x10*	=	grams

STEP 4

BAKE for	*20 x 2*	=	minutes

STEP 5

Serve warm with ice cream

Notes

Coconut Pie

STEP 1 Line a pie tin with sweet shortcrust pastry and blind bake until golden (check the packet for further instructions). Next reduce heat to 180°C

STEP 2

In a large bowl mix ingredients and pour into pie crust

MILK	*14+? = 5x3*	=	cup
EVAPORATED MILK	*2x ? =1*	=	cup
EGGS	*14-12*	=	
SHREDDED COCONUT	*1/3 of 3*	=	cup
SUGAR	*if I eat ¼ of a cake what fraction remains?*	=	cup
PLAIN FLOUR	*? x 6 =18*	=	Tbsp
BUTTER	*7+? = 4x2*	=	Tbsp

VANILLA ESSENCE

if I eat 4 pieces of pizza what fraction is left? = tsp

STEP 3

COOKING TIME

If jack is born in 1920 how old in 1970? = minutes

Chocolate Zucchini Cake

STEP 1 Preheat oven. Grease and line a 8 x 8" baking pan

TEMPERATURE	*50-10 +140*	=	°C

STEP 2

In a large bowl beat the eggs until light and fluffy. Whisk in zucchini and sugar

EGGS	*8÷4*	=	
ZUCCHINI, grated			
A pizza has 8 slices. How many pizzas do I need for 12 slices?		=	cups
SUGAR, **caster**	*Which is larger …. ⅕, ¼ or ½?*	=	cup

STEP 3

In a microwave safe bowl heat milk chocolate and coconut oil. Stop and occasionally stir until smooth. Add to the zucchini bowl and whisk.

DARK CHOCOLATE	*1000÷10*	=	grams
COCONUT OIL	*15minutes = ? of an hour (Hint: ⅓, ¼ or ½)*	=	cup

STEP 4

Add in flour, baking powder and salt. Mix to combine

WHEAT FLOUR	*60+60*	=	grams
BAKING POWDER	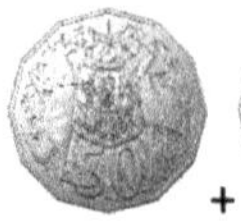 + 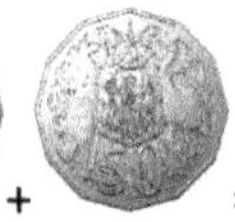 = how many dollars?	=	Tsp
SALT	*2x2-3*	=	pinch

STEP 5 Transfer batter to pan and bake

TIME	*¼ of 100*	=	minutes

STEP 6

Remove from oven and dust with icing sugar

Notes

Healthy Blueberry Cake

STEP 1 Pre-heat oven and grease 20 x20 cm pan

TEMPERATURE	*120+60*	=	°C

STEP 2

In a large bowl which together dry ingredients

OATS	*25x4*	=	grams
WHOLE WHEAT FLOUR	*25x6*	=	grams
BAKING POWDER	½ x3	=	tsp
BAKING SODA	¼ +¼	=	tsp
SALT	¼ +? = ½	=	tsp

STEP 3

In a medium bowl whisk together the following ingredients

GREEK YOGHURT	*3x3 = 8 +?*	=	cup
MILK	*What fraction of 8 = 4-2? (Hint: ⅓, ¼ or ½)*	=	cup
EGGS	*18 + ? = 10+10*	=	
VANILLA	*18½ + ? =20*	=	tsp
HONEY	*½ +6 +2 = 8+?*	=	cup

MELTED COCONUT OIL		*What fraction of the pizza is selected?*	=	cup

STEP 4

Add liquid ingredients to the dry ingredients and stir until just combined.

Gently fold blueberries into the mixture and bake

TIME	*22+23*	=	minutes
BLUEBERRIES	6÷3	=	cups

STEP 5

Cool in pan 20 minutes

Notes

No Bake Chocolate Slice

STEP 1

Grease and line an 20cm square pan with baking or parchment paper, ensuring two sides overhang.

STEP 2

Finely crush biscuits in a food processor and whiz

DIGESTIVE BISCUITS	*Half of 400*	=	grams

STEP 3

In a large mixing bowl, stir together crushed cookies with

RICE BUBBLES	*5 x 8*	=	grams
SHREDDED COCONUT	*5 x 6*	=	grams

STEP 3

Add into a mixing bowl. Heat in the microwave, stirring every 30 seconds, until smooth and creamy.

DARK CHOCOLATE	*¼ of 800*	=	grams

BUTTER UNSALTED	*If a plum weighs 30g how much do two plums weigh?*	=	grams
CONDENSED MILK	*24 x 10*	=	ml

STEP 4

Pour mixture over dry ingredients and mix quickly. When everything is coated with chocolate, pour into prepared tin and spread out into one layer. Place in the fridge for

TIME	*3 x (5+5)*	=	minutes

STEP 5

Add chocolate to a small bowl and melt in the microwave, stir every 30 seconds, until smooth.

MILK CHOCOLATE	*10 x 20*	=	grams

Pour over slice

STEP 6

Leave to set. Cut into small slices and serve

HEALTHY SCHOOL BALLS

Coconut Balls

STEP 1

Process the following ingredients until forms a mostly smooth paste

SHREDDED COCONUT	*16-4 = 11+?*	=	cup
COCONUT OIL, melted	*5x2 =12-?*	=	tsp

STEP 2

Add

MAPLE SYRUP	*1/3 of 9*	=	Tbsp
COCONUT MILK	*? is a ¼ of 8*	=	Tbsp
VANILLA	 *What fraction are red?*	=	tsp
CINNAMON	*4 is what fraction of 8? (Hint: ⅓, ¼ or ½)*	=	tsp
SALT	*13 = 12+?*	=	pinch

STEP 3

Add ½ cup of shredded coconut and pulse until just combined.

STEP 4

Shape the mixture into 12 (1-inch) balls and coat with the remaining shredded coconut and place on a plate. Refrigerate for at least 1 hour and up to 5 days.

OTHER FUN

Honey Joy Crackles

STEP 1 Preheat the oven

TEMPERATURE	*50x3*	=	°C

STEP 2 In a medium saucepan combine butter, honey & sugar over low heat until melted & frothy.

BUTTER	+ 35 50 =	=	grams
CASTER SUGAR	which is the largest ⅕, ¼ or ⅓?	=	cup

HONEY

If I can run 8 laps of the oval in 8 minutes. How many laps can I run in 1 minute?	=	Tbsp

STEP 3

Place cornflakes in a large bowl. Pour in the butter and sugar mixture. Fold gently until coated.

Spoon into patty pan holders and place on a flat baking tray.

CORNFLAKES	*13+3 = 4x?*	=	cups

STEP 4 Bake until golden.

TIME	*How many eggs in a dozen?*	=	minutes

STEP 5

Allow to cool & set before serving.

Baked Rice Pudding

Baked Rice Pudding

STEP 1

Preheat oven to	$\begin{array}{r} - \; 190 \\ 40 \\ \hline = \end{array}$	=	°C

STEP 2

In a large saucepan simmer the following ingredients over low heat. Stir frequently.

MILK, Full Cream	*500+400+100*	=	ml
ARBORIO RICE	*½ of 2 = ?*	=	cup(s)
SUGAR	*12-10 = ? x 1*	=	Tbsp
BUTTER	17= 19-?	=	Tbsp
SALT	 What fraction are apples?	=	tsp
COOKING TIME	*2x10 +10*	=	minutes

STEP 3

Remove from heat and stir in the raisins, vanilla and whisked eggs.

RAISINS	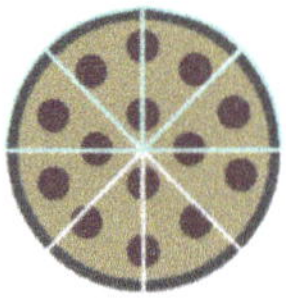 If I eat 4 slices, what fraction remains?	=	cup
EGGS	*A dozen minus 10*	=	
VANILLA	*6x? = 12*	=	tsp

STEP 4 Pour the rice pudding into a ceramic baking pan. Drizzle the condensed milk all over the top.

CONDENSED MILK	*1- ½*	=	cup
COOKING TIME	*15+15*	=	minutes

PEANUT BUTTER CHOC CHIP COOKIES

STEP 1:		
PRE-HEAT OVEN	= 190	°C
STEP 2:		
BUTTER SOFTENED	= ½	CUP
BROWN SUGAR	= 1	CUP
STEP 3:		
EGG	= 1	
PEANUT BUTTER	= ½	CUP
VANILLA EXTRACT	= 2	TSP
STEP 4:		
PLAIN FLOUR	= 1 ½	CUPS
BAKING SODA	= ½	TSP
SALT	= ½	TSP
STEP 5:		
CHOCOLATE CHIPS	= 2	CUPS
STEP 7:		
COOKING TIME`	= 15	MINUTES

COCONUT MERINGUE COOKIES

STEP 1:		
TEMPERATURE	= 160	°C
STEP 2:		
EGG WHITES	= 3	
CREAM OF TARTAR	= ¼	TSP
CASTER SUGAR	= ¾	CUP
STEP 3:		
COCONUT FLAKES	= 1	CUP
CORNFLAKE CRUMBS	= ¾	CUP
STEP 5:		
COOKING TIME	=18	MINUTES

SEED CRACKER

STEP 1:		
FLOUR	= ½	CUP
PEPITA/PUMPKIN SEEDS	=1	CUP
CHIA SEED	= ¼	CUP
SEA SALT	= 1	TSPN
OLIVE OIL	= 2	TBSP
WATER	= 250	ML
STEP 2:		
OVEN	= 180	°C
STEP 3:		
THICKNESS	= 3	MM
STEP 5:		
BAKE	=1	HOUR

CINNAMON APPLE MUFFIN

STEP 1:		
TEMPERATURE	= 180	°C
STEP 2:		
PLAIN FLOUR	= 2	CUPS
BAKING POWDER	= 1 ½	TSP
CINNAMON	= 2 ½	TSP
SALT	= 1	PINCH
STEP 3:		
PINK LADY APPLES	= 2	CUPS
STEP 4:		
BUTTER	= ¼	CUP
CASTER SUGAR	= 2/3	CUPS
EGG	= 2	
VANILLA	= 2	TSP
STEP 5:		
MILK	= ½	CUP
STEP 6:		
TIME	= 30	MINUTES
STEP 7:		
CINNAMON	= 3	TBSP
SUGAR	= ¼	CUP
STEP 8:		
BUTTER, MELTED	= ¼	CUP

BANANA & CHOC CHIP MUFFIN

STEP 1:		
TEMPERATURE	= 200	°C
STEP 2:		
WHOLMEAL FLOUR	= 2	CUPS
BAKING POWDER	= 3	TSP
BROWN SUGAR	= 1/3	CUP
STEP 3:		
EGGS	= 1	
OLIVE OIL	= ½	CUP
MILK	= 1	CUP
STEP 4:		
BANANA	= 1	CUP
CHOC CHIPS	= 1/2	CUP

CHOCOLATE AVOCADO FUDGE MUFFIN

STEP 1		
TEMPERATURE	= 160	°C
STEP 2:		
FLOUR	= 2	CUPS
SUGAR	= ¾	CUP
COCOA	= ½	CUP
SALT	= ½	TSP
CHOCOLATE CHIPS	= ¾	CUP DARK
BAKING POWDER	=1	TSP
STEP 3:		
MASHED AVOCADO	= 1	
GREEK YOGHURT	= 150	G
SKIM MILK	= ONE	CUP
COCONUT OIL	= 1/4	CUP
VANILLA	= 1	TSP
STEP 5: BAKE		
APPROXIMATE TIME	=40	MINUTES

FETA & ROAST CAPSICUM MUFFIN

STEP 1:		
TEMPERATURE	= 200	°C
STEP 2:		
WHOLEMEAL FLOUR	= ½	CUPS
BAKING POWDER	= 3	TSP
STEP 3:		
EGGS	= 1	
OLIVE OIL	= ½	CUP
MILK	= 1	CUP
STEP 4:		
ROAST CAPSICUM	= 2/3	CUP
FETA, CRUMBLED	= ½	CUP
STEP 6:		
TIME	= 20	MIN

SAVOURY RICOTTA MUFFIN

STEP 1:		
TEMPERATURE:	=190	°C
STEP 2:		
SUN DRIED TOMATOES	= 4	
OLIVES	= 2	TBSP
CAPERS	= 2	TSP
STEP 3:		
EGGS	= 2	
RICOTTA	= 1	CUP
SPRING ONION	= 2	
SALT & PEPPER	= 1	PINCH EACH
STEP 5:		
APPROXIMATE TIME	= 20	MIN

OMELETTE SANDWICH

STEP 1:		
OIL	=1	TSP
CAPSICUM	= 10	G
RED ONION	= ½	
BABY SPINACH LEAVES	= 1	HANDFUL
BABY TOMATOES	= 6	
STEP 2:		
EGGS	=4	
MILK	= 2	TBSP

WHITE BEAN AND CHEESE SANDWICH

STEP 1:		
CANNELINI BEANS	=1	TIN
OLIVE OIL	=1	TBSP
LEMON JUICE	=1 ½	TBSP
SALT	= 1	PINCH
PEPPER	=1	PINCH
STEP 3:		
FETA CHEESE	= ¼	CUP

APPLE WEETBIX PIKELETS

STEP 1:		
WHOLEMEAL FLOUR	= 1	CUP
CRUSHED WEETBIX	= 1 ½	CUPS
BAKING POWDER	= 2	TSP
CINNAMON	=2	TSP
STEP 2:		
EGG	=1	
BUTTER	=2	TBSP
MILK	=1 ½	CUPS

TUNA SLICE

STEP 1:		
TEMPERATURE	= 180	°C
STEP 2:		
DRY CRACKERS	= ½	PACKET
ONION , BROWN	= 1	
CHEESE, CHEDDAR	= 1 ½	CUPS
STEP 3:		
EGGS	= 6	
MILK	= ½	CUP
TUNA	= 2	TINS
FROZEN VEGETABLES	= 2	CUPS
ZUCCHINI	= 1	
STEP 4:		
TIME	= 30	MINUTES

SLOW COOKER YELLOW BEEF CURRY

STEP 1:		
BEEF	= 500	G
POTATOES, DICED	= 4	
ONION DICED	= 1	
RED CAPSICUM DICED	= 1	
COCONUT CREAM	= 400	ML CAN
BEEF STOCK	= ½	CUP
GINGER	= 1	TBSP
YELLOW THAI CURRY PASTE	= 3	TBSP
GARLIC CRUSHED	= 3	TSP
FISH SAUCE	= 1	TBSP
BROWN SUGAR	= 1	TBSP
STEP 2:		
TIME	= 8	HOURS
STEP 3:		
CORN FLOUR	= 1	TBSP
CURRY SAUCE	= 2	TBSP
APPROXIMATE TIME	= 30`	MINUTES
STEP 4:		
JUICE	= 1	TBSP

TUNA AND VEGETABLE ONE POT DUMP BAKE

STEP 1:		
TEMPERATURE	= 180	°C
STEP 2:		
ONION, BROWN	= 1	
FROZEN VEGE MIX	= 1	KG BAG
SHREDDED CHEESE	= 2	CUPS
TUNA CAN, LARGE	= 425	G TIN
SOUP	= 420	G TIN
SOUR CREAM	= 2/3	CUP
BLACK PEPPER	= ½	TSP
STEP 3:		
TIME	= 30	MINUTES
STEP 4:		
TIME	= 5	MINUTES

SAAG PANEER

STEP 2:		
SPINACH	= 1	PACKET
STEP 3:		
PANEER, CUBED	= 600	GRAMS
STEP 4		
CUMIN SEEDS	= 1	TSP
OLIVE OIL	= 1	TBSP
STEP 5:		
TIME	= 2	MINUTES
GARLIC, MINCED	= 2	CLOVES
RED ONION	= 1	CUP
CANOLA OIL	= 1	TBSP
STEP 6:		
PUREED TOMATO	= 1	TIN
SALT	= 1	TSP
TIME	= 5	MINUTES
STEP 7:		
WATER	= 2	CUP
CURRY POWDER	= 1	TBSP
CORIANDER POWDER	= 1	TSP
CHILI POWDER	= 1 ½	TSP
STEP 8:		
CREAM	= 2	TBSP
BUTTER	= 1	TBSP
TIME	= 5	MINUTES

SHREDDED BRUSSELS SPROUTS WITH PISTACHIOS, CRANBERRIES & PARMESAN

STEP 2:		
BROWN ONION	= 1	
TIME	= 5	MINUTES
OLIVE OIL	= 1	TSP
STEP 3:		
OLIVE OIL	= 1	TSP
BRUSSEL SPROUTS	= 500	G
STEP 4:		
PISTACHIO	= 1/3	CUP
DRIED CRANBERRIES	= 1/3	CUP
PARMESAN CHEESE	= 1/3	CUP
PEPPER	= 1	TSP

HOMEMADE PESTO PASTA / PASTA SALAD

STEP 1:.		
SPIRAL PASTA	=2	CUPS
STEP 2:		
BASIL LEAVES, FRESH	= 2	CUPS
PINE NUTS	= ½	CUP
STEP 3:		
GARLIC	= 1	CLOVE
PARMESAN CHEESE	= ½	CUP
STEP 4:		
OLIVE OIL	= ¼	CUP
SALT	= ½	TSP
PEPPER	= ½	TSP
STEP 5:		

SUNDRIED TOMATOES	= ½	CUP

PUMPKIN ROASTED HAZELNUTS AND FETA SALAD

STEP 1:		
TEMPERATURE	=200	°C.
TIME	=10	MINUTES
HAZELNUT	= 1	CUP
STEP 3:		
PUMPKIN	= 1.5	KG
SALT	= 1	TSP
PEPPER	= 1	TSP
TIME	= 15	MINUTES
STEP 4:		
HONEY	= ¾	CUP
OLIVE OIL	= 2	TBSP
BALSAMIC VINEGAR	= 2	TBSP
STEP 5:		
FETA CHEESE	=175	G

BLUEBERRY CHEESECAKE

STEP 2:		
DIGESTIVE BISCUITS	= 180	G
TIME	= 30	SECONDS
STEP 3:		
UNSALTED BUTTER	= 80	G
STEP 4:		
TIME	= 20	MIN
STEP 5:		
CREAM CHEESE	= 300	G
DOUBLE CREAM	= 250	ML
LEMON JUICE	= 4	TBSP
VANILLA EXTRACT	= 1	TSP

BLUEBERRY DUMP CAKE

STEP 1:		
PREHEAT OVEN TO	= 180	°C
STEP 2:		
BLUEBERRIES	= 3	CUPS
SUGAR	= ½	CUP
CORNFLOUR	= 2 ½	TBSP
SALT	= 1	PINCH
STEP 2:		
PLAIN FLOUR:	= 175	G
SUGAR:	= 200	G
BAKING POWDER	= 1	TSP
SALT:	= ¼	TSP
STEP 3:		
BUTTER, MELTED	= 170	G
STEP 4:		
BAKE FOR	= 40	MINUTES

COCONUT PIE

STEP 2:		
MILK	= 1	CUP
EVAPORATED MILK	= ½	CUP
EGGS	= 2	
SHREDDED COCONUT	= 1	CUP
SUGAR	= ¾	CUP
PLAIN FLOUR	= 3	TBSP
BUTTER	= 1	TBSP
VANILLA ESSENCE	= ½	TSP
STEP 3:		
TIME	= 50	MINUTES

CHOCOLATE ZUCCHINI CAKE

STEP 1:		
TEMPERATURE	= 180	°C
STEP 2:		
EGGS	= 2	
ZUCCHINI, GRATED	= 1 ½	CUPS
SUGAR, CASTER	= ½	CUP
STEP 3:		
DARK CHOCOLATE	= 100	G
COCONUT OIL	= ¼	CUP
STEP 4:		
FLOUR	= 120	G
BAKING POWDER	= 1	TSP
SALT	= 1	PINCH
STEP 5:		
APPROXIMATE TIME	=25	MINUTES.

HEALTHY BLUEBERRY CAKE

STEP 1:		
TEMPERATURE	= 180	°C
STEP 2:		
OATS	= 100	G
WHOLE WHEAT FLOUR	= 150	G
BAKING POWDER	=1 ½	TSP
BAKING SODA	= ½	TSP
SALT	= ¼	TSP
STEP 3:		
GREEK YOGHURT	=` 1	CUP
MILK	= ¼	CUP
EGGS	= 2	
VANILLA	= 1 ½	TSP
HONEY	= ½	CUP
MELTED COCONUT OIL	= 1/3	CUP
STEP 4:		
APPROXIMATE TIME	= 45	MINUTES
BLUEBERRIES	= 2	CUPS

NO BAKE CHOCOLATE SLICE

STEP 2:		
DIGESTIVE BISCUITS	= 200	G
STEP 3:		
RICE BUBBLES	= 40	G
SHREDDED COCONUT.	= 30	G
STEP 3:		
DARK CHOCOLATE	= 200	G
BUTTER UNSALTED	= 60	G
CONDENSED MILK	= 240	ML
STEP 4:		
APPROXIMATE TIME	= 30	MIN
MILK CHOCOLATE	= 200	G

COCONUT BALLS

STEP 1:		
SHREDDED COCONUT	= 1	CUP
COCONUT OIL	= 2	TSP
STEP 2:		
MAPLE SYRUP	= 3	TBSP
COCONUT MILK	= 2	TBSP
VANILLA	= ½	TSP
CINNAMON	= ½	TSP
SALT	= 1	PINCH

HONEY JOY CRACKLES

STEP 1:		
TEMPERATURE	= 150	°C.
STEP 2:		
BUTTER	= 85	G
CASTER SUGAR	= 1/3	CUP
HONEY	= 1	TBSP
STEP 3		
CORNFLAKES	= 4	CUPS
STEP 4:		
TIME	= 12	MIN

BAKED RICE PUDDING

STEP 1:		
TEMPERATURE	= 150	°C
STEP 2:		
MILK, FULL CREAM	= 1000	ML
ARBORIO RICE	= 1	CUP(S)
SUGAR	= 2	TBSP
BUTTER	= 2	TBSP
SALT	= 1/3	TSP
TIME	= 30	MIN
STEP 3:		
RAISINS	= ½	CUP
EGGS	= 2	
VANILLA	= 2	TSP
STEP 4		
CONDENSED MILK	= ½	CUP
TIME	= 30	MIN

GET YOUR NAME IN THE NEXT EDITION!

We want your photographs of our recipes so that we can include your work in the next edition. So take a close up photo of your work using a phone or camera (remember to clean the lens).

COPY THIS FORM AND GET A PARENT OR GUARDIAN TO SIGN.

Forms can be returned via email to *learningthroughfun1@gmail.com*

This photograph is the property of ______________________ and I give my permission for it to be republished.

Photograph by ______________________

City ______________________

Country ______________________

Recipe name ______________________

Signature ______________________

Role (ie Guardian/Parent) ______________________

IF OUR TEAM LOVES YOUR PHOTO WE WILL INCLUDE IT IN THE NEXT PRINT AND YOU WILL BE NOTIFIED BY EMAIL

Dear readers,

Thank you for taking the time to read our book. As a small independent company your support immensely helps us and encourages us to pursue new work. We would be grateful if you can leave a review on Amazon, Goodreads or any other forum that you think would be helpful in spreading the word and helping other readers decide whether or not to read the book. Don't forget to email us any feedback at *learningthroughfun1@gmail.com*.

Perhaps you have a family recipe you wish to share with the world. If so please follow the following format or email and we will send details. We may need to adapt the recipe for regional ingredient availability but will try and stay true to the recipe you love. We look forward to seeing your name and recipe in future editions!

Name ____________________
City ____________________
Country ____________________
Age ____________________
Recipe ____________________

Please provide an ingredient list and step by step instructions. Be guided by this book format.

Publications in this Range

Learning Through Fun - Maths Year 2

Learning Through Fun - Maths Year 3

Learning Through Fun - Maths Year 4

Learning Through Fun - Maths Year 5

www.ingramcontent.com/pod-product-compliance
Lightning Source LLC
LaVergne TN
LVHW070857210926
848482LV00016B/963
* 9 7 8 0 6 4 8 6 0 8 7 0 7 *